Geometry and Fractions with Tangrams

Grades 3-6

Barbara Bando Irvin, Ph.D.

Copyright © 1995 Learning Resources, Inc., Vernon Hills, Illinois (U.S.A.)
Learning Resources Ltd., King's Lynn, Norfolk (U.K.)

ISBN: 1-56911-972-4

Printed in China

Contents

Introduction

Geometry and Fractions with Tangrams contains 43 blackline master activities to help students in grades 3-6 reinforce basic mathematical concepts of geometry and fractions, apply problem-solving strategies, and increase spatial visualization skills. The activities may be expanded to accommodate and enrich your mathematics curriculum.

The activities in this book are designed to be used with a set of tangrams. A set of overhead tangrams may be used to demonstrate various tangram activities and problems on the overhead projector, or may be used by students to show a particular solution to their classmates. A blackline master of two sets of tangrams is provided on page 8. Color and cut out several copies of this page so that students can make large geometric shapes as well as whole pages of patterns. Selected solutions and an Award Certificate can be found at the back of the book.

Overview of the Content

The activities in this book were developed using NCTM's *Curriculum and Evaluation Standards for School Mathematics* (1989) and *Geometry and Spatial Sense, Addenda Series K-6* (1993) as guidelines. The mathematical content in this book begins by concentrating on geometric concepts, then blends in basic fraction concepts. Specific goals are:

- To identify, describe, and model geometric shapes.
- To combine tangram pieces to make and describe other geometric shapes.
- To measure, describe, and classify angles.
- To explore congruence and similarity of shapes.
- To recognize symmetric shapes and locate lines of symmetry.
- To measure area using nonstandard units of measure.

The understanding of fractional number relationships and the operations of addition with fractional numbers are also investigated. Specific goals are:

- To name fractional parts of a region.
- To name and represent unit and proper fractions.
- To add and subtract fractions.
- To multiply and divide fractions.
- To find perimeter and area using standard units of measure.

Tangram activities are designed to accomplish spatial visualization skills. Specific goals are:

- To improve visual discrimination.
- To improve visual memory.
- To improve eye-motor perception.
- To improve figure-ground perception.
- To develop spatial and proportional reasoning skills.
- To improve position-in-space perception.

Using Geometry and Fractions with Tangrams

Each section begins with *Teaching Notes* to provide an overview of the content and suggestions for classroom use. The *Teaching Notes* include:

Getting Ready

An activity to introduce the mathematical concepts in the section.

Activity Teaching Notes

The objective and a description of each blackline master activity is provided to help you integrate it into your mathematics curriculum.

Tangram Activities

Blackline masters for each activity are described in the *Teaching Notes*.

Students should be encouraged to work together cooperatively in pairs or small groups to do an activity or to find all possible solutions. Urge students to create their own problems and puzzles to challenge each other. Tangrams are an enjoyable, hands-on, motivational approach to work with an array of mathematical concepts, artistic relationships, and creations.

Place several sets of tangrams in a learning center for students to manipulate during free time or to complete various tangram activities. You may wish to copy and laminate several of the blackline master activities in this book. Provide several levels of tangram puzzles for students to solve, including shapes with interior lines, outlines of shapes, and silhouettes of figures. Challenge students by providing tangram puzzles that use two sets of tangrams or feature tangram paradoxes.

Problem-Solving and Visualization Skills

Students can sharpen their problem-solving and spatial visualization skills with the geometric and fractional concepts on the activity pages. Experiences with tangrams help students sharpen spatial sense by teaching them to recognize shapes in different positions and combinations of shapes. The relationship among the tangram pieces and the decision of the placement of tangram pieces to fill in an outline or silhouette is also beneficial for students' conceptual development. Encourage students to make as many geometric shapes as possible such as rectangles, pentagons, and irregular hexagons. To help students enhance their visualization skills, tangrams can be used to create designs and "pictures." Some examples below show a cat, a bird, a candle, a house, the letter E, and the numeral 2. Begin with figures that contain interior lines, then progress to outlines of figures, and finally to silhouettes of figures.

Getting Started

Allow time for students to familiarize themselves with the set of tangrams. Ask them about the number of sides for each piece. Observe and listen to students as they manipulate the pieces. Some students may discover that two small triangles can cover the medium triangle, a few others may make "animal" or "people" figures, while others form geometric shapes. Ask students about their discoveries. Share them with their classmates. Ask students about the different shapes (*triangle, square, parallelogram*) in a tangram set. Review other geometric shapes including *rectangle, hexagon,* and *trapezoid.*

About Tangrams

The tangram is an ancient Chinese puzzle consisting of seven pieces in three different shapes: two large triangles, one medium triangle, two small triangles, a square, and a parallelogram. The large triangle is twice the area of the medium triangle. The medium triangle, the square, and the parallelogram are each twice the area of a small triangle. Each angle of the square measures 90°. Since each triangle contains a 90° angle and two 45° angles, they are all isosceles right triangles, and the two sides opposite the 45° angles are congruent. The parallelogram contains 45° and 135° angles. The relationships among the pieces enables them to fit together to form many figures and arrangements.

Constructing a Tangram

Students can gain valuable experience learning and reviewing various geometric concepts and vocabulary terms by constructing the tangram puzzle through paper folding. The materials needed are a model of the tangram square (page 8), a 4-inch square piece of paper, and a pair of scissors. To emphasize geometric concepts, ask the questions below during the construction process.

1. Show students the square pattern with internal shape lines, and the 4-inch square piece of paper.

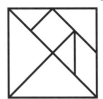

- How would you begin to fold and cut the square piece of paper to form the seven pieces of the tangram? What would be your first fold? [A diagonal of the square.]

2. Fold the square piece of paper in half along the diagonal. Cut along the diagonal fold to make two triangles.

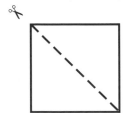

- Which side of the diagonal in the pattern is easier to work with in order to form some of the tangram pieces? [The part showing two triangles.]

3. Fold one of the triangles in half. Cut along the fold line to make two smaller triangles. Label them "1" and "2."

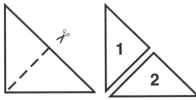

- Are the two triangles alike? [Yes.]
- Can you fit one triangle on top of the other? [Yes.]
- Are they *congruent*? [Yes.]
- How do you know? [Each triangle is the same size and shape.]
- What do you notice about the angles of these triangles? [A right angle and two smaller angles of the same size.]
- Can you name this triangle more accurately? [A *right triangle*.]
- What is the long side of each triangle called? [The *hypotenuse*.]
- Are the shorter sides of each triangle the same size? [Yes.]

A triangle having at least two congruent sides is *isosceles*.

- Can you name this triangle even more accurately? [An *isosceles right triangle*.]

Ask older students:

- How could you figure out the size of the smaller angles of the triangle? [Place the two triangles together so that the smaller (45°) angles form a right angle. Half of a right angle (90°) is 45°. Triangles 1 and 2 will form the two large triangles of the tangram.]

4. Now work with the remaining half piece of paper to form the other five pieces of the tangram. Fold the triangle in half and then open it up to show a fold line.

- Where is the hypotenuse of this triangle? [The long side opposite the right angle.]
- What is the relationship between the fold line and the hypotenuse? [They are *perpendicular*; they form right angles.]

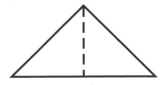

5. Take the vertex of the right angle of the triangle and fold it down to the midpoint of the hypotenuse. (The vertex of the right angle should meet at the intersection of the hypotenuse and the first fold line.) Cut along the new fold line.

- What two shapes are formed? [A triangle and a trapezoid.]

 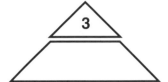

Label this triangle "3."

- How is triangle 3 similar to triangles 1 and 2? [It has a right angle.]
- How is triangle 3 different from triangles 1 and 2? [It's smaller.]
- Are the shapes the same? [Yes.] Figures having the same shape with corresponding congruent angles are *similar*. Triangle 3 is the medium triangle of the tangram.

6. Show the trapezoid with the longer side at the bottom.

- Does it contain any right angles? [No.]
- Are any of the angles congruent to any other angles? [Yes.]
- How do you know? [The smaller angles of shapes 1, 2, and 3 fit over the smaller angles of the trapezoid.] Now fold one lower corner of the trapezoid so that it falls on the midpoint of the longest side of the trapezoid. Crease it and then cut along the fold line.

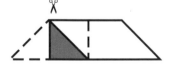 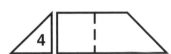

- What is the shape of the piece that is cut off? [A triangle.] Label it "4."
- What does the remaining shape look like? [A trapezoid with a right angle.] Cut off the square along the fold line and label it "5."

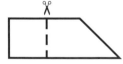 

- What kind of angles are in the square? [Right angles.]

- What does the final remaining shape look like? [A trapezoid with a right angle.]

- Are there any other types of angles in this shape? [*Acute* and *obtuse* angles.]

7. Fold the vertex of the right angle up to the vertex of the obtuse angle. Crease it. Cut along the fold line.

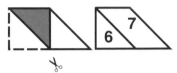

- What two pieces are formed? [A triangle and a *parallelogram*.]

8. Label the triangle "6" and the parallelogram "7."

- Are there any right angles in the parallelogram? [No.]
- What kinds of angles are in the parallelogram? [Acute and obtuse angles.]

Ask older students:

- Can you figure out how many degrees in the acute angle? [It's the same size as the smaller angle in the right triangle which is 45°.]
- Can you figure out the measure of the obtuse angle? [Since a right angle and a 45° can fit into the obtuse angle, its measure is 90° + 45° = 135°.]
- What can you say about triangles 4 and 6? [Same size and shape; they're *congruent*.]

- What can you say about triangles 3 and 4? [Same shape but different size; they're *similar*.]

- What can you say about triangles 1 and 4? [Same shape but different size; they're *similar*.]

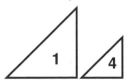

Geometry and Fractions with Tangrams
©Learning Resources, Inc.

Tangram Pieces and Other Shapes

Vocabulary

triangle, square, parallelogram, rectangle, hexagon, trapezoid, acute, right, obtuse, convex, concave, pentagon

Getting Ready

Allow time for students to familiarize themselves with the set of tangrams. Ask them about the number of sides for each piece. Observe and listen to students as they manipulate the pieces. Some students may discover that two small triangles can cover the medium triangle, others may make "animal" or "people" figures, while others may form geometric shapes. Ask students about their discoveries and to share them with the classmates. Ask about the different shapes (*triangle, square, parallelogram*) of the tangram pieces. Review other geometric shapes including *rectangle, hexagon*, and *trapezoid*.

Tangram Activities

The Tangram Pieces (*pages* 10-11)

The questions on these two pages focus on the types of polygons (*triangles, quadrilaterals*).

Their descriptions in terms of sides and angles should be assigned for use at one time. Discuss the shape and name of each tangram piece. This is a good summary for the **Getting Ready** activity.

Tangram Angles (*pages* 12-13)

The measures for *acute, right*, and *obtuse* angles are given at the top of page 12. Encourage students to fit the corners of each tangram piece into the angles shown to categorize and measure each corner angle of the tangrams. Ask students how they determined that the acute angles in the tangram pieces are 45° and the obtuse angles are 135°. Also note that the sum of the angles of a triangle is 180° and the sum of the angles of a quadrilateral is 360°.

Other Geometric Shapes; Convex and Concave Shapes (*pages* 14-19)

Before assigning any of the activities, students should have some idea how the sides of different tangram pieces fit together. Have students find and fit together any two tangram pieces so that one side of a tangram piece fits exactly along the side of another tangram piece. Also ask them how many different ways two pieces can fit together to make different shapes (see examples below).

Ask students questions similar to those below about two-piece combinations:

- Can any of the sides of a large triangle and a medium triangle fit together? [Yes]
- Can any of the sides of a large triangle and the square fit together? [No]
- Can any of the sides of the parallelogram and the square fit together? [Yes]

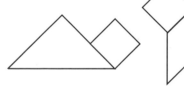

The activities on pages 14-19 concentrate on two objectives: Convex and concave polygons and making other geometric shapes. A *convex* polygon has no indentations. A *concave* polygon contains an indention (or "cave") where it is possible to draw a line segment between two points of the region such that part of the segment lies outside the polygonal region. Polygons that will be formed on these pages include *squares, parallelograms, triangles, rectangles, trapezoids, hexagons, and pentagons*. Have students save their activities on pages 10-19 in order to complete a polygon chart shown in the *Teaching Note*s on page 21.

Convex   *Concave*

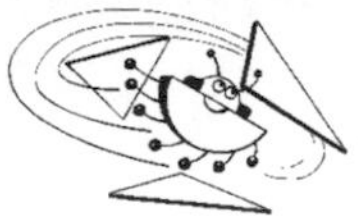

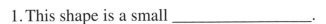

The Tangram Pieces

Cover each figure below with the appropriate tangram piece. Then answer each question with words or numbers.

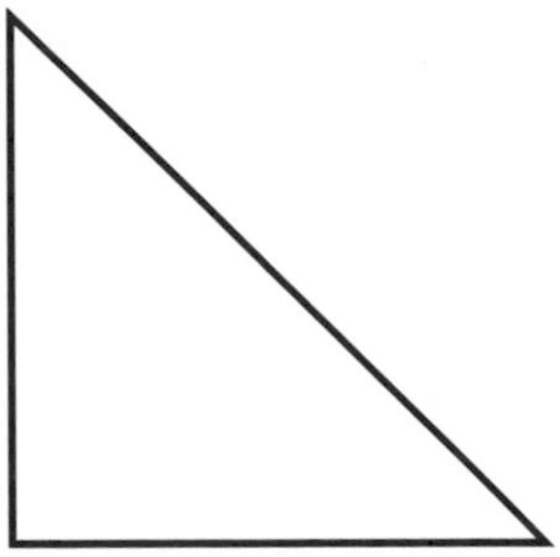

1. This shape is a small _____.

   It has _____ sides and _____ angles.

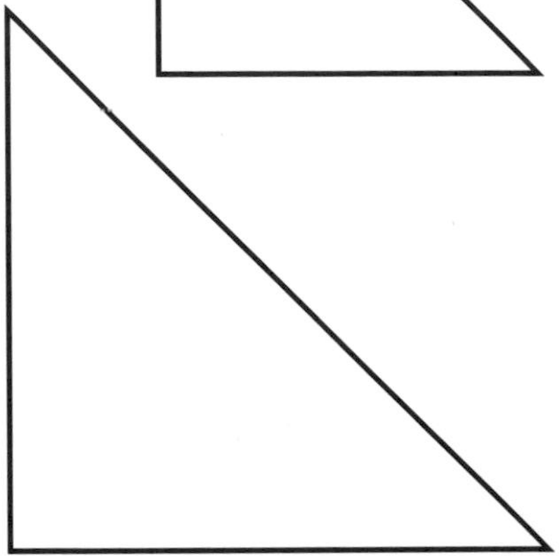

2. This shape is a medium _____.

   It has _____ sides and _____ angles.

   How many small triangles can cover this shape? _____

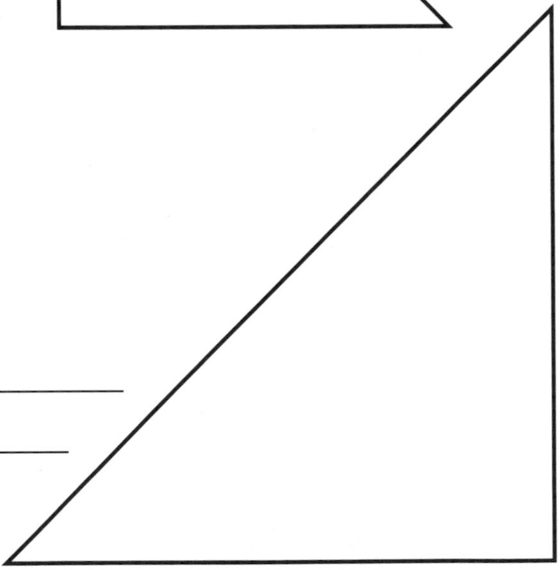

3. This shape is a large _____.

   It has _____ sides and _____ angles.

   How many medium triangles can cover this shape? _____

   How many small triangles can cover this shape? _____

4. How are the small, medium, and large triangles the same? _____

   _____

5. How are the small, medium, and large triangles different? _____

   _____

Geometry and Fractions with Tangrams
©Learning Resources, Inc.

More Tangram Pieces

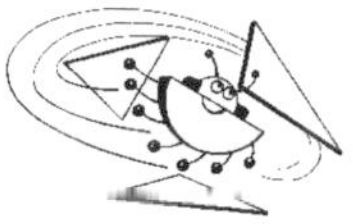

Cover each figure below with the appropriate tangram piece. Answer each question with words or numbers.

1. This shape is a_____.

   It has_____sides and_____angles.

   All the sides are_____.

   How many small triangles can cover this shape?_____

2. This shape is a_____.

   It has_____sides and_____angles.

   How many small triangles can cover this shape?_____

3. How are the square and parallelogram the same? _____

   _____

4. How are the square and parallelogram different?_____

   _____

5. The_____and the_____above are quadrilaterals.

6. A quadrilateral has_____sides and_____angles.

7. Circle the figures below that are quadrilaterals.

A. 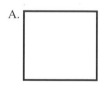   B.   C.

D.   E.   F.

Angles of Tangram Pieces

Here are three types of angles:

Right Angle	Acute Angle	Obtuse Angle	Turns
Measures 90 degrees.	Measures between 0 and 90 degrees.	Measures between 90 and 180 degrees.	A complete turn is 360°.
90°	**45°**	**135°**	**360°**

1. Fit each corner of the square piece in the right angle above. Label them. What is the measure of each angle?_____

2. Fit each corner of the triangles below in the angles above. Label them.

   The large angle is a_____angle.

   The smaller angles are_____angles.

3. Fit each corner of the parallelogram in the angles above. Label them. Are there any right angles in the parallelogram? _____

   The smaller angles are_____angles.

   The larger angles are_____angles.

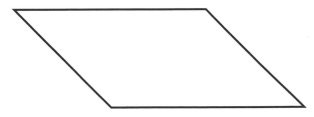

4. What is the sum of all the angle measures of a triangle?_____

5. What is the sum of all the angle measures of a quadrilateral?_____

Geometry and Fractions with Tangrams
©Learning Resources, Inc.

More About Angles and Sides

Cover each figure below with the appropriate tangram piece. Answer each question with words or numbers.

1. This shape is a_____.

   It is a quadrilateral that has_____sides and_____angles.

   All the sides are_____.

   It could be called a rectangle with_____sides.

   It could be called a rhombus with_____angles.

   Label each angle with the appropriate number of degrees.

2. This shape is a _____.

   It is a _____ with 4 sides and 4 angles.

   It has two angles with a measure _____ than 90° and

   two angles with a measure _____ than 90°.

   Label each angle with the appropriate number of degrees.

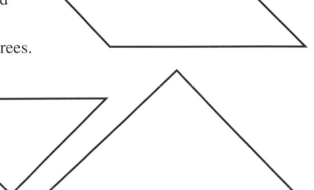

3. These shapes are _____.

   Each of them has _____sides and

   _____ angles.

   There is a _____ angle in each triangle.

   There are two _____ angles in each triangle.

   A special name for a triangle containing a right angle

   is a _____ _____.

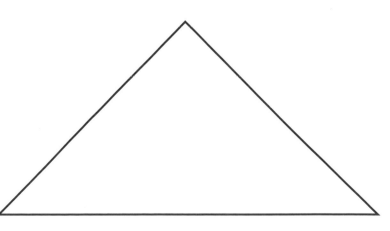

Convex and Concave Polygons

Two or more tangram pieces can be put together to make other shapes. The shapes are *convex* or *concave*. Concave polygons have a "cave."

Example: Use a medium and small triangle to make two different shapes.

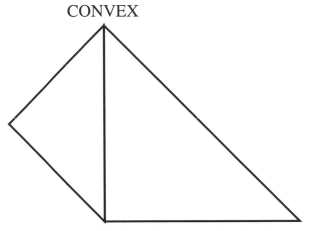

CONVEX   CONCAVE

1. Use two small triangles to make the three convex polygons named below.

1 SQUARE   1 TRIANGLE   1 PARALLELOGRAM

2. Use two small triangles to make at least two concave polygons.

Geometry and Fractions with Tangrams
©Learning Resources, Inc.

Convex and Concave

1. Use a medium triangle and a parallelogram to make a convex polygon and a concave polygon.

2. Use two small triangles and a parallelogram to make the three convex polygons named below.

RECTANGLE   PARALLELOGRAM   TRIANGLE

3. Make at least three different concave polygons.

Convex and Concave

1. Use the small triangle
   and the square to make
   one convex shape
   and one concave
   shape.

2. Use a medium triangle, two small triangles, and a square to make three convex shapes and
   two concave shapes.

More Convex Polygons

1. Use two small triangles and a square to make the four convex polygons named below.

        RECTANGLE   PARALLELOGRAM

        TRAPEZOID   TRIANGLE

2. Use two small triangles and a parallelogram to make four different convex polygons below.

More Convex Polygons

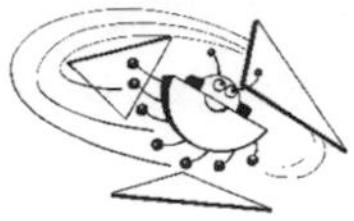

Use a medium triangle, two small triangles, and a square to make the following convex polygons:

RECTANGLE

PARALLELOGRAM

TRAPEZOID

PENTAGON   HEXAGON

Seven-Piece Convex Polygons

Use all seven pieces to make each convex polygon.

RECTANGLE

TRAPEZOID

PARALLELOGRAM

Congruent and Similar Shapes

Vocabulary

congruent, similar

Getting Ready

Have students look at the square. Ask them to make the square another way using other tangram pieces.

- Are both squares the same shape? [Yes.]
- Are both squares the same size? [Yes.]

Encourage students to look at the medium triangle. Ask them to make the medium triangle another way using other tangram pieces.

- Are both triangles the same shape? [Yes.]
- Are both triangles the same size? [Yes.]

Finally, have students look at the large triangle. Ask them to make the large triangle another way using other tangram pieces.

- Are both triangles the same shape? [Yes.]
- Are both triangles the same size? [Yes.] Each pair of squares, medium triangles, and large triangles are *congruent* because they have the same size and shape.

Tangram Activities

Congruent Shapes (*pages* 22-23)

On page 22, students use three or four tangram pieces to make rectangles. Rectangles A and B are congruent because they have the same size and shape. Ask students to use three other tangram pieces to make another rectangle congruent to A or B. [Use two small triangles and one medium triangle.] Rectangle C is a different rectangular shape than A or B. And, rectangle D is similar to A and B. On page 23, congruent parallelograms and squares are constructed. Ask students whether they can make another parallelogram congruent to the two they constructed. [Use one large triangle, one parallelogram, and two small triangles.]

Similar Shapes (*page* 24)

When two figures are the same shape and proportional in size, they are *similar*. Corresponding angles of similar triangles are congruent. The problems on page 24 compare three squares. Ask students to use more tangram pieces to make larger squares and then ask if any of the squares are similar to each other. [Yes.]

Congruent and Similar Shapes (*pages* 25-30)

On pages 25-30, students are asked to make various polygons using a specified number of tangram pieces. After each set of parallelograms (page 25), triangles (page 26 and 29), rectangles (pages 27 and 28), and hexagons (page 30) have been constructed, a set of completion or true/false statements are listed to help students compare the figures. Discuss these statements with students. You may wish to extend these exercises by asking students to construct pentagons using three or more tangram pieces and then comparing them for congruence or similarity. Have students save these pages in order to complete the polygon chart shown on page 21.

Polygon Chart

Make a large chart on the bulletin board for students to complete as they finish each activity sheet in this book. Challenge them to make polygons using one to seven tangram pieces.

Number of Pieces / Polygon	1	2	3	4	5	6	7
Square	□	S/S M/M					
Triangle							
Parallelogram							
Rectangle			S/S				
Trapezoid							
Pentagon							
Hexagon							

Have students complete a chart making polygon shapes using two sets of tangrams.
Ask if they can make an octagon or a nonagon.

Same-Size Shapes

1. Use three different tangram shapes to make two rectangles. Make the second one congruent to the first one.

2. Use a parallelogram, a medium and two small triangles to make a rectangle.

3. Use a square, a medium, and two small triangles to make a rectangle.
   Are the two rectangles congruent? _____

Geometry and Fractions with Tangrams
©Learning Resources, Inc.

Congruent Shapes

1. Use a square, a large, and two small triangles to make a parallelogram

   Then use a large, a medium, and two small triangles to make a parallelogram.

   Are the two parallelograms congruent? _____

2. Use all seven tangram pieces to make two congruent squares.

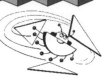

Same Shape, Different Size

The triangles below have the same shape but are different sizes. They are *similar* polygons. Place the small triangle on top of the medium triangle at each corner. What do you notice?

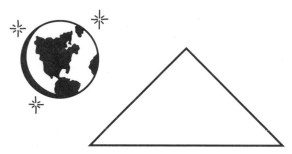

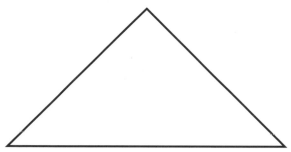

1. Use two small triangles to make a square. Label it A. Then use two large triangles to make a square. Label it B. Are the squares *similar* or *congruent*? _____

2. Use a medium and two small triangles to make a square next to the other two squares above. Label it C. Is this square *similar* or *congruent* to the two other squares? _____

3. Use a large and two small triangles and a square to make a square.

   Is this square *congruent* to any of the squares above. _____

   Which ones? _____

   Is this square *similar* to any of the squares above? _____

   Which ones? _____

Congruent or Similar Trapezoids

1. Use two small triangles and a square to make a trapezoid. Label it A.

2. Use a medium and two small triangles to make a trapezoid. Label it B.

3. Use a large and two small triangles and a parallelogram to make a trapezoid. Label it C.

4. Write *congruent* or *similar* in the blanks to compare trapezoids A, B, and C above.

   a. Trapezoid A is _____ to Trapezoid B.

   b. Trapezoid B is _____ to Trapezoid C.

   c. Trapezoid A is _____ to Trapezoid C.

Congruent or Similar Triangles

1. Use a medium and two small triangles to make a triangle. Label it A.

2. Use two small triangles and a square to make a triangle. Label it B.

3. Use a large and two small triangles and a square to make a triangle. Label it C.

4. Use a large and two small triangles and a parallelogram to make a triangle. Label it D.

5. Write *congruent* or *similar* in each blank to compare the triangles above.

   a. Triangle A is _____ to Triangle C.   d. Triangle A is _____ to Triangle D.

   b. Triangle B is _____ to Triangle A.   e. Triangle C is _____ to Triangle B.

   c. Triangle C is _____ to Triangle D.   f. Triangle D is _____ to Triangle B.

Geometry and Fractions with Tangrams
©Learning Resources, Inc.

Congruent or Similar Rectangles

1. Use two small triangles and a square to make a rectangle. Label it A.

2. Use two small triangles and a parallelogram to make a rectangle, Label it B.

3. Use a medium triangle, two small triangles, and a parallelogram to make a rectangle. Label it C.

4. Use a large, a medium, and two small triangles to make a rectangle. Label it D.

5. Write true or false (T or F) for each statement.

   a. Rectangle B is congruent to Rectangle A. _____

   b. Rectangle C is similar to Rectangle A. _____

   c. Rectangle D is similar to Rectangle A. _____

   d. Rectangle B is congruent to Rectangle C. _____

   e. Rectangle B is similar to Rectangle C. _____

   f. Rectangle B is congruent to Rectangle D. _____

Comparing More Rectangles

1. Make three different rectangles using five tangram pieces for each one.
   Draw them below and label them A, B, and C.

2. Are any of the rectangles congruent to each other? _____ Which ones? _____

3. Are any of the rectangles similar to each other? _____ Which ones? _____

Make Large Triangles

1. Use four tangram pieces to make a triangle.

2. Use seven tangram pieces to make a triangle.

3. Are the triangles congruent to each other?_____ Explain your answer._____

   _____

4. Are the triangles similar to each other?_____ Explain your answer._____

   _____

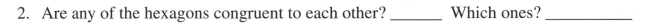

Make Hexagons

1. Make four different hexagons using four tangram pieces for each one.
   Label them A, B, C, and D. Use different combinations of pieces in each hexagon.

2. Are any of the hexagons congruent to each other? _____ Which ones? _____

3. Are any of the hexagons similar to each other? _____ Which ones? _____

4. Are any of the hexagons regular hexagons? _____ Which ones? _____

Symmetry of Tangram Shapes

Vocabulary

lines of symmetry

Getting Ready

Direct students to find shapes in the classroom that are symmetric. Shapes are symmetric when one half of the shape can be folded over the other half exactly. Ask how many ways a shape can be folded in half in this manner. The fold lines are *lines of symmetry*. Students will discover that shapes can have no lines of symmetry, only one line of symmetry, and several lines of symmetry. Ask students how many lines of symmetry there may be in a leaf, T-shirt, or paper plate.

In this section of activities, students will not only discover which shapes are symmetric, they will also review the concepts of convex and concave shapes.

Tangram Activities

Symmetry of Convex Shapes (*pages* 32-34)

On page 32, students are to determine which tangram pieces are symmetric and how many lines of symmetry are contained in those that are symmetric. Provide scissors for the cut-and-fold activities.

Students are to use four tangram pieces to make the polygons listed. All of the polygons have at least one line of symmetry except the parallelogram. The rectangle and the hexagon each have two lines of symmetry. Extend the lesson by selecting a different set of four tangram pieces to have students construct other symmetric polygons.

The rectangle in problem 1 and the hexagon in problem 3 each have two lines of symmetry. The quadrilateral constructed for problem 2 is a square with four lines of symmetry. After completing pages 33 and 34, ask students to think about the following questions:

- Do all squares have four lines of symmetry?
- Do all rectangles have two lines of symmetry?
- Do all hexagons have two lines of symmetry?
- Are all pentagons symmetric?
- Are all trapezoids symmetric?

Extend the activities on pages 33 and 34 by asking students to construct convex symmetric shapes using five, six, and seven tangram pieces.

Symmetry of Concave Shapes (*pages* 35-36)

On pages 35 and 36, students can make various shapes that resemble single-headed arrows, double-headed arrows, houses, people, or birds to make concave shapes with one or two lines of symmetry. Here are some sample solutions:

Shapes with 1 line of symmetry:

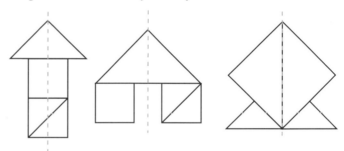

Shapes with 2 lines of symmetry:

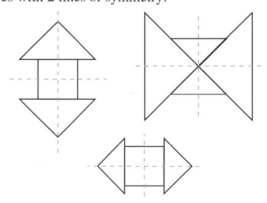

Extend the activities on pages 35 and 36 by asking students to construct concave symmetric shapes using five, six, and seven tangram pieces.

Symmetry

1. Cut out each tangram piece at the bottom of the page.

2. Fold each piece in half as many ways as you can.

3. Unfold each piece and place it on its identical tangram shape shown below.

4. Use a dotted line to draw the fold line of each piece.

5. Write the number of lines of symmetry on each tangram piece.

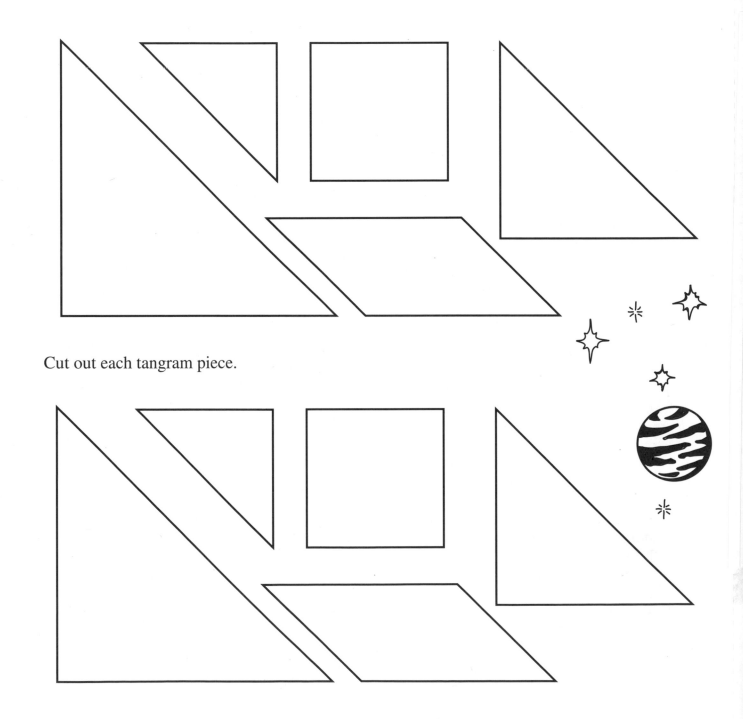

Cut out each tangram piece.

Name..Date.................

Make Symmetric Shapes

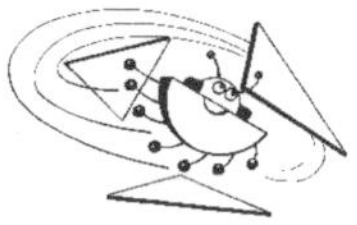

Use a medium, two small triangles, and a parallelogram to make each polygon symmetric. Trace each shape and use dotted lines to draw the lines of symmetry.

RECTANGLE

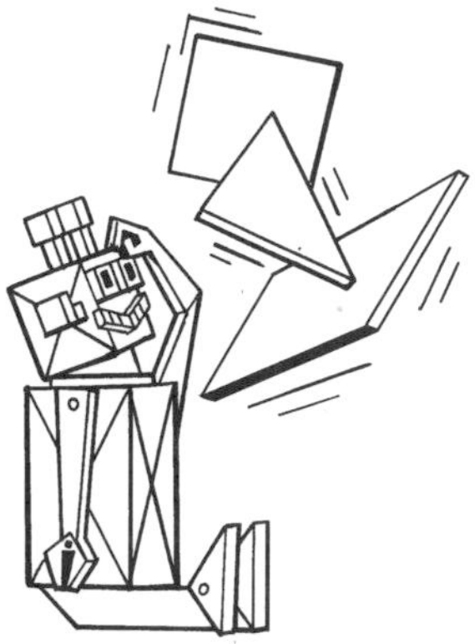

PARALLELOGRAM

TRAPEZOID

HEXAGON

PENTAGON

Many Lines of Symmetry

1. Use two large and two small triangles and a parallelogram to make a quadrilateral with two lines of symmetry.

2. Use a medium triangle, two small triangles, a parallelogram, and a square to make a quadrilateral with four lines of symmetry.

3. Use two small triangles, a square, and a parallelogram to make a hexagon with two lines of symmetry.

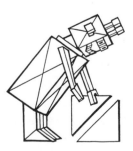

Symmetric Concave Shapes

Use any combination of four tangram pieces to make at least four concave shapes with one line of symmetry.

More Symmetric Concave Shapes

Use any combination of three or four tangram pieces to make at least four con-cave shapes with two lines of symmetry.

Geometry and Fractions with Tangrams
©Learning Resources, Inc.

Area Measurement with Nonstandard Units

Vocabulary

area

Getting Ready

Students in grades 3-6 should be familiar with "covering" or "tiling" activities as a method to approach the concept of area. When given an outlined region and units such as pattern block squares or cubes to cover the entire region, they can cover the region and then count the number of units used. Sometimes, this procedure includes parts of units (halves and quarters) in order to make a very good approximation for area. Use pattern block squares or 3 x 5 index cards to cover the front of a book or the top of a desk to explore the area concept.

Encourage students to make areas with a specific number of units. For example, using pattern block or tangram triangles, have students make regions with 4 or 15 units.

Tangram Activities

Area and Nonstandard Units (*pages* 38-40)

On page 38, the tangram square is used as a unit of area. For problems 3, 5, and 6, students must think in half units in order to find the area of each shape. Extend the lesson by asking students to construct shapes with areas of 8 units, $2\frac{1}{2}$ units, and $5\frac{1}{2}$ units.

The small triangle is used on pages 39 and 40 as a unit of area. However, this tangram piece represents two different units of area. After students have found the area of each shape on page 39, ask them to find the area using the square or the medium triangle as a unit of area. Then compare the number of units of area for each figure using the smaller and larger units. If the unit of area doubles in size, then the number given for the area is half the number it would be if the smaller unit were used.

On page 40, the small triangle represents a $\frac{1}{2}$ unit of area. After completing this page, ask students to make tangram shapes with $2\frac{1}{2}$, 4, and $5\frac{1}{2}$ units of area.

Area and Money Values (*pages* 41-42)

The activities on pages 41 and 42 involve large amounts of money. Area is given in terms of monetary values. On page 41, the small triangle has the least value and each of the other larger shapes have a greater value. On page 42, the medium triangle is given a value. Have students find the value of each tangram piece in the upper left corner of the page before they do the problems. Encourage students to make a set of Tangram Money Cards with a value chart and a polygon on each card. Allow them to challenge each other to find the values, make the shapes, or both. Display the cards on a bulletin board or make them available in the learning center of the room.

Measuring with Square Units

If a square = 1 unit of area, then what is the area of each shape?

1.

Area = _____ units

1 Unit

2.

Area = _____ units

3.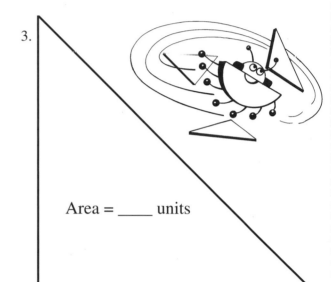

Area = _____ units

4.

Area = _____ units

5.

Area = _____ units

6.

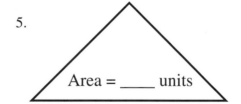

Area = _____ units

Measuring Areas

If a small triangle = 1 unit of area, then what is the area of each shape?

1.

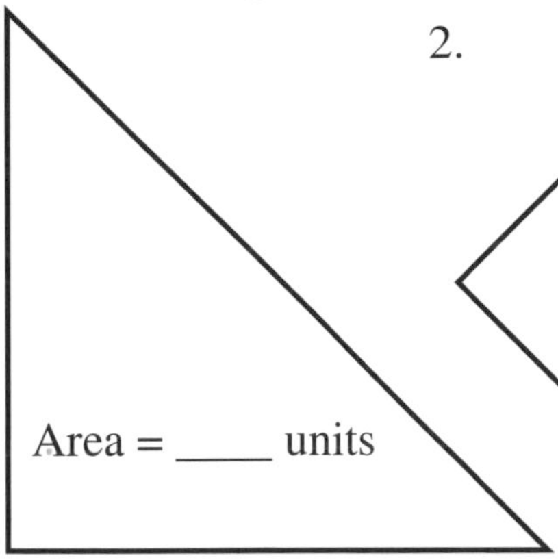

Area = _____ units

2.

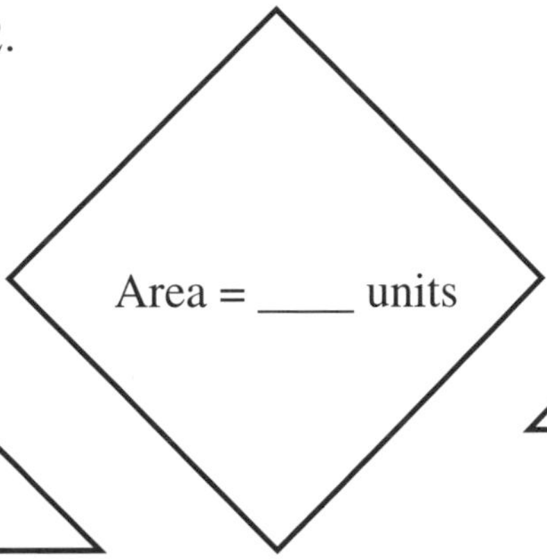

Area = _____ units

3.

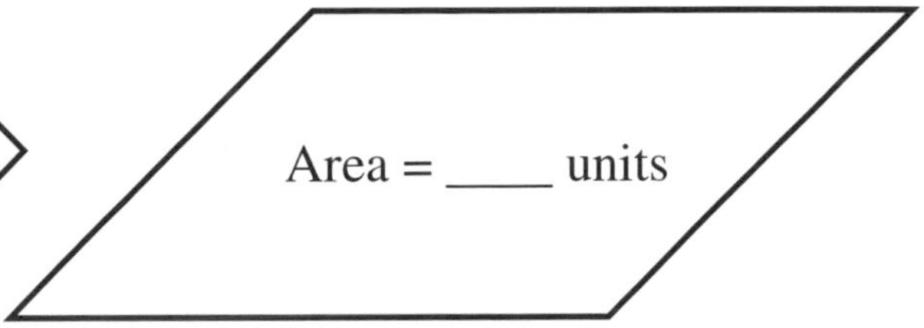

Area = _____ units

4.

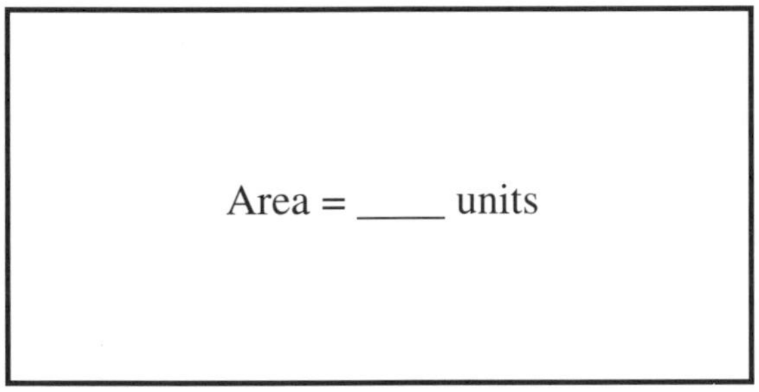

Area = _____ units

5.

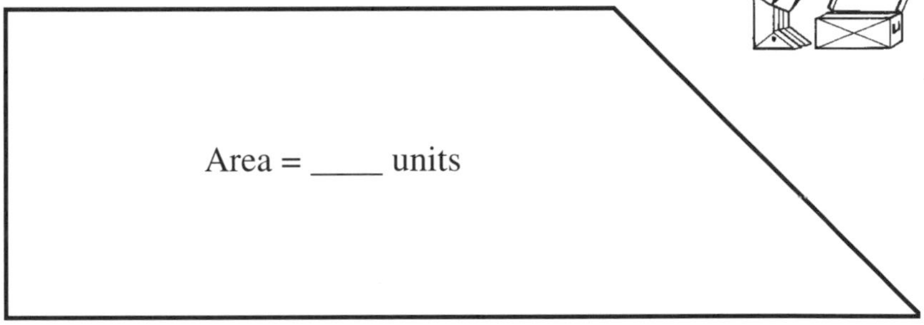

Area = _____ units

6.

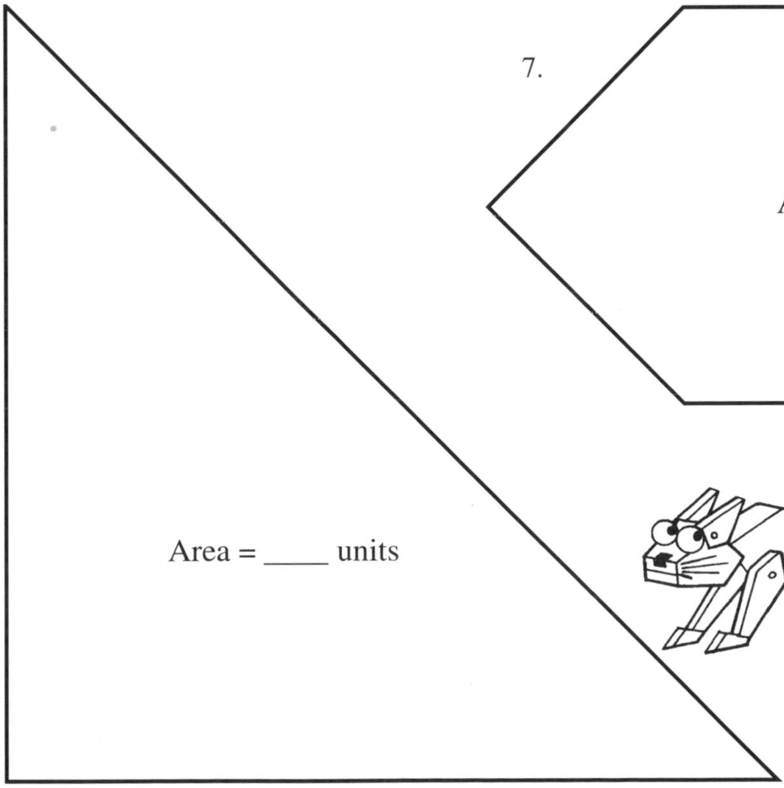

Area = _____ units

7.

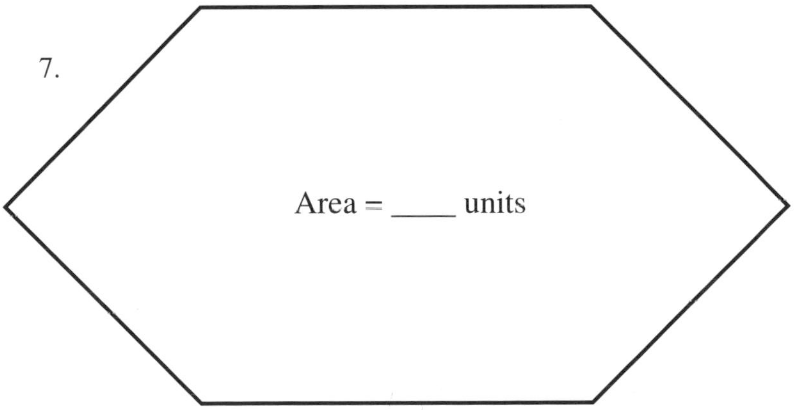

Area = _____ units

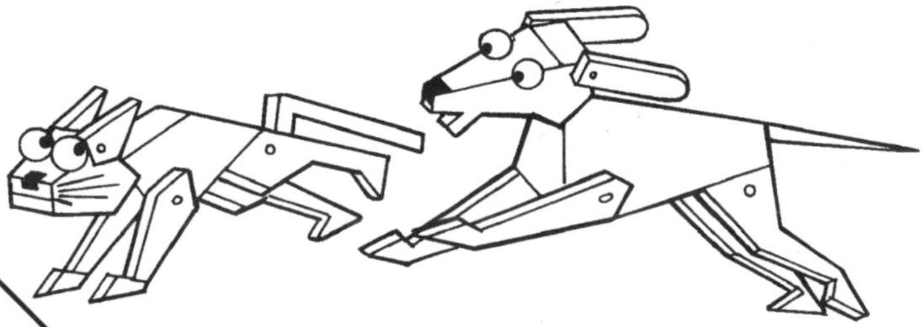

Make Areas

Make each shape with the least number of tangram pieces. Trace each piece. If a small triangle is $\frac{1}{2}$ unit of area, then what is the area of each shape?

1.

Area = _____ units

2.

Area = _____ units

3.

Area = _____ units

4.

Area = _____ units

5.

Area = _____ units

6.

Area = _____ units

Geometry and Fractions with Tangrams
©Learning Resources, Inc.

Big Bucks!

If a small triangle = $250, then
Medium triangle = $ _____
Large triangle = $ _____
Square = $ _____
Parallelogram = $ _____
What is the value of all
 seven tangram pieces? _____

1. Make this pentagon with the least number of tangram pieces. What is the value of this pentagon?

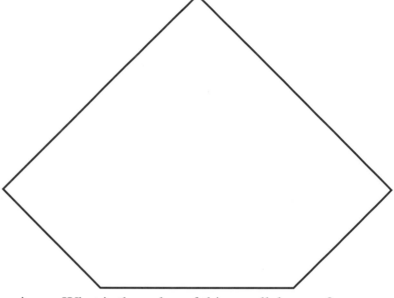

2. Make this parallelogram with tangram pieces. What is the value of this parallelogram? _____

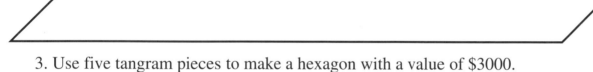

3. Use five tangram pieces to make a hexagon with a value of $3000.

More Big Bucks!

If a medium triangle = $1000, then

Small triangle = $ _____

Large triangle = $ _____

Square = $ _____

Parallelogram = $ _____

What is the value of all
seven tangram pieces? _____

1. Use five tangram pieces to make this square.
What is the value of this square?

2. Make this trapezoid with tangram pieces. What is the value of this trapezoid? _____

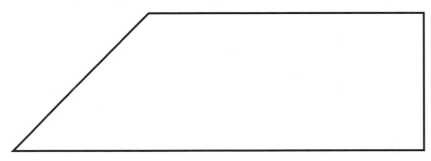

3. Use six tangram pieces to make a rectangle with a value of $6000.

Geometry and Fractions with Tangrams
©Learning Resources, Inc.

Fractional Parts of Tangram Shapes

Vocabulary

fraction, parts

Getting Ready

Have students look at a complete tangram puzzle (page 8). Review the area relationships between all the triangles, the triangles and the square, and parallelogram. For example, the large triangle has twice the area of the medium triangle. Or, two small triangles can cover the parallelogram. Also ask how much of the whole tangram puzzle is covered by the two large triangles [$\frac{1}{2}$].

Tangram Activities

Tangram Piece Fractional Parts (*page* 44)

Encourage students to figure out the fractional value of each tangram piece and record it on page 44. Finding the total sum of the fractional parts of the tangram will give students an informal introduction for adding fractions with unlike denominators. Although seven addends are necessary to find the sum, this fraction addition problem is not overwhelming for students since they can visualize the sum of all the tangram pieces by looking at the whole tangram puzzle. Direct students to write the fraction value of each tangram piece in the equation provided on page 44, then have them express the entire equation with a common denominator ($\frac{1}{16}$).

$$\frac{1}{4} + \frac{1}{4} + \frac{1}{8} + \frac{1}{16} + \frac{1}{16} + \frac{1}{8} + \frac{1}{8} = 1$$

$$\frac{4}{16} + \frac{4}{16} + \frac{2}{16} + \frac{1}{16} + \frac{1}{16} + \frac{2}{16} + \frac{2}{16} = \frac{16}{16} = 1$$

Fractional Parts of Shapes (*pages* 45-49)

The exercises on pages 45-49 use the same values as those shown on page 44. Permit students to use page 44 as a guide to help them solve the problems on these pages.

On pages 45 and 46, students are to write an addition equation for each tangram shape in order to find the sum. Note that the interior lines are shown in each shape on page 45 and only the shape outlines are given on page 46. The addition equations may vary on page 46.

Page 47 presents students with the challenge of making tangram shapes that satisfy a subtraction equation. Several solutions are possible for problems 1 and 4.

These problems are very similar to the area problems with nonstandard units encountered earlier on pages 39 and 40, except the value of the small triangle is now $\frac{1}{16}$. Finding the solution to each problem requires the iteration of the small triangle on each shape. The equations reflect this iteration by multiplying the number of small triangles times $\frac{1}{16}$ to find the product. You may wish to expand this lesson by creating problems that use a pattern block or a fraction square piece as the unit of area.

Tangrams can be used to informally explore notions of division with fractions due to the relationship between the areas of each tangram piece. After students complete the problems, discuss them and then have students draw tangram pictures to verify their solutions for problems 4-10. Problem 10 is especially challenging in that $\frac{3}{8}$ is not divisible by $\frac{1}{4}$ a whole number of times [$\frac{3}{8} \div \frac{1}{4} = 1\frac{1}{2}$].

Fraction Challenges (*page* 50)

Page 50 involves fraction values for each tangram piece that are different from those established on page 44. Before assigning page 50, present students with problems such as the one shown below.

If this shape = 1, what is the value of each tangram piece?

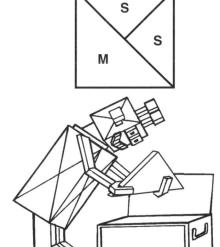

Tangram Parts

Place all seven of the tangram pieces on this square. Move them around and compare the pieces to find the fractional part of each piece.

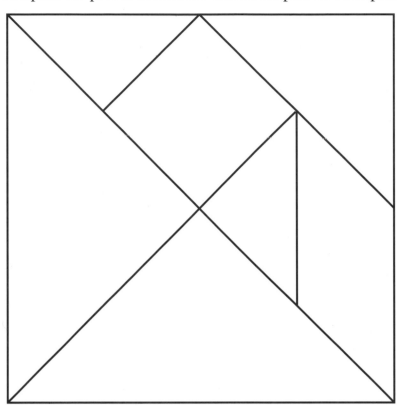

1. If the area of the large square is 1, what is the area of :

   Large triangle = _____

   Medium triangle =_____

   Small triangle =_____

   Square =_____

   Parallelogram =_____

Label each tangram piece with its fraction value.

2. Add the tangram pieces.

Large triangle + Large triangle + Medium triangle + Small triangle + Small triangle + Square + Parallelogram.

_____ + _____ + _____ + _____ + _____ + _____ + _____ = _____

Look at the tangram puzzle above.

3. What fractional part of the whole puzzle are the two large triangles?_____

4. What fractional part of the whole puzzle is a large triangle?_____

5. What fractional part of a large triangle is a medium triangle?_____

6. What fractional part of the whole puzzle is a medium triangle?_____

7. What fractional part of the whole puzzle is a small triangle?_____

8. What fractional part of the whole puzzle is the square?_____

9. What fractional part of the whole puzzle are all the triangles ?_____

Geometry and Fractions with Tangrams
©Learning Resources, Inc.

Fractional Values of Shapes

Using small triangle $= \frac{1}{16}$, find the fractional value of each shape. Write an addition equation under each shape to find the sum.

1.

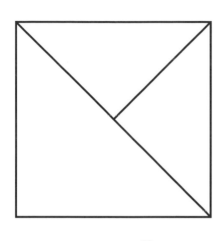

_____ = _____

2.

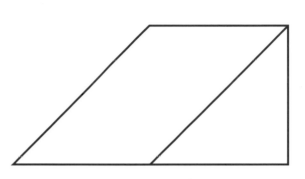

_____ = _____

3.

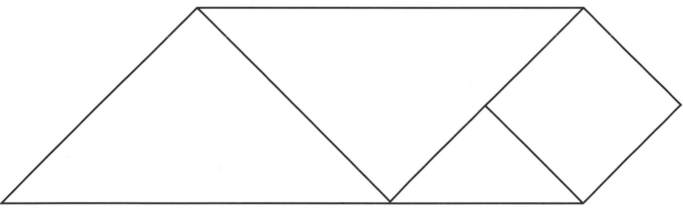

_____ = _____

4.

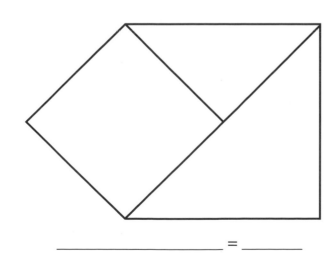

_____ = _____

Geometry and Fractions with Tangrams
©Learning Resources, Inc.

More Fractional Values of Shapes

Using large triangle = $\frac{1}{4}$, find the fractional value of each shape. Place and trace the appropriate tangram pieces on each shape. Then write an addition equation under each shape to find the sum.

1.

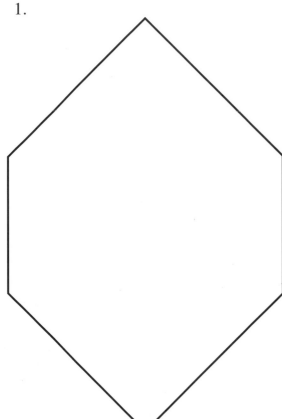

_____ = _____

2.

_____ = _____

3.

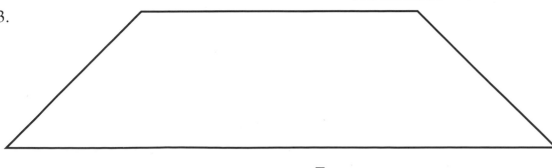

_____ = _____

Geometry and Fractions with Tangrams
©Learning Resources, Inc.

Missing Fraction Parts

Using medium triangle = $\frac{1}{8}$, make convex shapes to show the solution for each subtraction equation. Place and trace each piece in the shapes. Label each piece with a fraction.

1. Use three pieces to show $1 - \frac{3}{4} = $ _____   2. Make a 4-piece hexagon to show $1 - \frac{5}{8} = $ _____

3. Use five pieces to make a trapezoid to show $1 - \frac{1}{2} = $ _____

4. Use five pieces to make a parallelogram to show $1 - \frac{1}{4} = $ _____

Multiply Fraction Parts

If a small triangle = $\frac{1}{16}$, find the value of each shape. Place and trace small triangles on each shape and complete the multiplication equations. Write the product in lowest terms.

1.

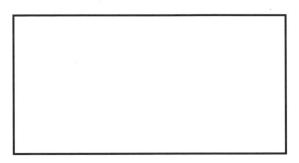

_____ x $\frac{1}{16}$ = _____

2.

_____ x $\frac{1}{16}$ = _____

3.

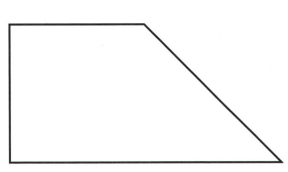

_____ x $\frac{1}{16}$ = _____

4.

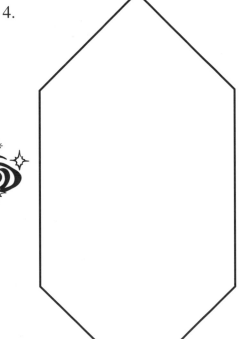

5.

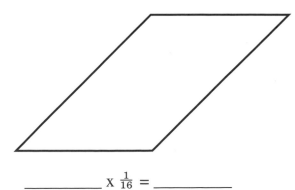

_____ x $\frac{1}{16}$ = _____

_____ x $\frac{1}{16}$ = _____

Geometry and Fractions with Tangrams
©Learning Resources, Inc.

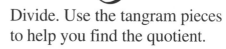

Parts of Parts

Use a small triangle $= \frac{1}{16}$, for each problem. Place and trace the appropriate tangram pieces on each shape to find the answer to each division problem.

1. How many small triangles are in this shape?

_____ ÷ _____ $\frac{1}{16}$ = _____

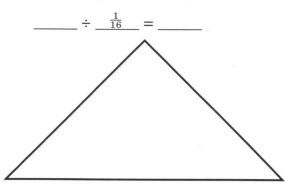

Divide. Use the tangram pieces to help you find the quotient.

4. $\frac{1}{4} \div \frac{1}{16} =$ _____

5. $\frac{1}{8} \div \frac{1}{8} =$ _____

6. $\frac{1}{2} \div \frac{1}{4} =$ _____

7. $\frac{2}{4} \div \frac{1}{8} =$ _____

2. How many medium triangles are in this shape?

_____ ÷ _____ $\frac{1}{8}$ = _____

8. $\frac{3}{8} \div \frac{1}{16} =$ _____

9. $\frac{3}{4} \div \frac{1}{8} =$ _____

10. $\frac{3}{8} \div \frac{1}{4} =$ _____

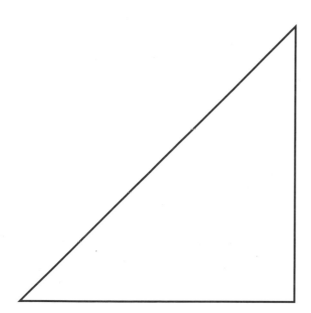

3. How many small triangles are in this shape?

_____ ÷ _____ = _____

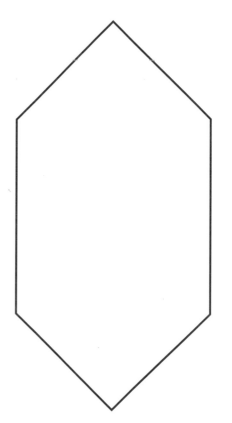

Geometry and Fractions with Tangrams
©Learning Resources, Inc.

Fraction Challenges

1. If this shape has a value of 1, what is the value of each piece?

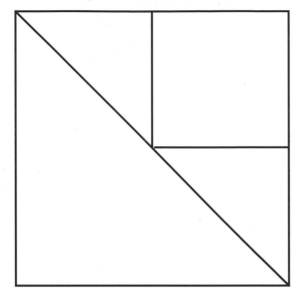

Large triangle = _____

Medium triangle = _____

Small triangle = _____

Square = _____

Parallelogram = _____

2. If this shape has a value of 1, what is the value of each piece?

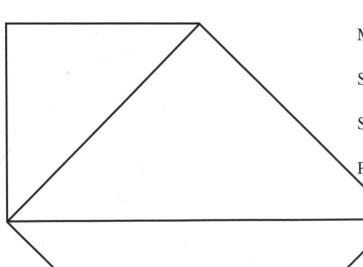

Large triangle = _____

Medium triangle = _____

Small triangle = ___

Square = _____

Parallelogram = _____

Geometry and Fractions with Tangrams
©Learning Resources, Inc.

Area and Perimeter with Standard Units

Vocabulary

area, perimeter, units, square units

Getting Ready

Direct students to find the perimeter of a classroom object such as the cover of their mathematics book using small paper clips. Ask them how they found the distance around the cover. Then have them measure the lengths of each side of the book cover with an inch ruler. Since a small paper clip is approximately one inch long, the perimeter of the book cover in terms of paper clips and inches are about the same amount. Repeat this measurement activity using centimeter units.

Tangram Activities

Area using Inches (*pages 52-53*)

Use pages 52 and 53 together so that students have the inch grid and the table of area values handy to complete problems on both pages. By arranging each piece on the inch grid in special ways, students can find the area of each tangram piece to the nearest whole number. The exact amount of area can be determined two ways: (1) by counting the number of whole and half-squares and (2) by using logical reasoning based on the relationships between the tangram pieces. For example, if the large triangle has an area of 4 square inches, then the area of the medium triangle is 2 square inches, and the area of the small triangle is 1 square inch. Since the square and parallelogram can be constructed using two small triangles, the area of each is 2 square inches.

Area using Centimeters (*pages 54-55*)

Pages 54 and 55 should also be used together. By arranging each piece on the centimeter grid, students can figure out the area of each piece to the nearest $\frac{1}{2}$ and $\frac{1}{4}$ centimeter by counting and estimating the number of centimeter squares. Students can use logical reasoning to find the exact area. If the large triangle has an area of 25 sq. cm, then the medium triangle, parallelogram, and square each have an area of $12\frac{1}{2}$ (12.5) sq. cm, and the small triangle has an area of $6\frac{1}{4}$ (6.25) sq. cm.

Students may express the areas as mixed numbers or decimals. Finding the areas of various polygonal shapes on these two pages involves the addition of mixed numbers or decimals. Insist that students write the addition problems for each shape. Extend the lessons on pages 52-55 by having students create area problems to challenge each other.

Perimeter using Inches (*pages 56-57*)

Finding the perimeter of the sides of the tangram pieces can easily be determined when the set of tangram pieces is placed on the 4 x 4-inch grid. Using the grid, students can see that the hypotenuse of the large triangle is 4 inches, the hypotenuse of the small triangle and the longer side of the parallelogram are each 2 inches. However, the interior lines of the tangram puzzle may be determined by measuring with an inch ruler or using the square root of 2 since the ratio for the lengths of the sides for a 45-45-90 right triangle is 1:1: $\sqrt{2}$. Measuring to the nearest $\frac{1}{8}$ inch, the lengths of the sides are 4 inches, $2\frac{3}{4}$ inches, 2 inches, and $1\frac{3}{8}$ inches. Using $\sqrt{2}$, to find the hypotenuse of the medium triangle, multiply 2 x $\sqrt{2}$ (or 1.414...) which is about 2.8 inches. Since a side of the square and the shorter side of the parallelogram is half the length of the length of the hypotenuse of the medium triangle, its length is about 1.4 inches. Measuring these lengths to the nearest $\frac{1}{8}$ inch comes very close to the calculated lengths; compare $2\frac{3}{4}$ (2.75) inches with 2.82 ... and $1\frac{3}{8}$ (1.375) inches with 1.41....

Perimeter using Centimeters (*pages 58-59*)

Students can figure out the dimensions of each tangram in centimeters the same as was done for inches. Measuring to the nearest half centimeter, the lengths of the sides are 10 cm, 7 cm, 5 cm, and 3.5 cm. Calculating using $\sqrt{2}$, the lengths of the sides are 10 cm, 7.07 cm, 5 cm, and 3.53 cm.

Inches of Area

1. Each square on the grid is 1 square inch. Arrange the tangram pieces on the grid to find the area of each piece in square inches (sq. in.).

Large triangle = _____ sq. in.

Medium triangle = _____ sq. in.

Small triangle = _____ sq. in.

Square = _____ sq. in.

Parallelogram = _____ sq. in.

The area of the whole tangram puzzle is _____ sq. in.

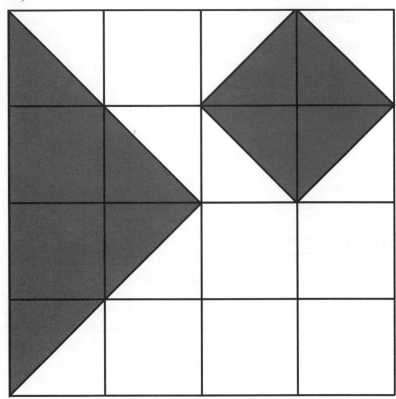

2. Use three tangram pieces to make a trapezoid with an area of 4 square inches.

3. Use four tangram pieces to make a rectangle with an area of 6 square inches.

Geometry and Fractions with Tangrams
©Learning Resources, Inc.

Inches of Area

1. Use five tangram pieces to make a trapezoid with an area of 12 square inches.

2. Use six tangram pieces to make a pentagon with an area of 14 squares inches.

3. Use five tangram pieces to make a convex polygon with an area of 13 square inches.

Centimeters of Area

1. Each square on the grid is 1 square centimeter. Arrange the tangram pieces on the centimeter grid to find the area of each piece in square centimeters (sq. cm or cm^2).

   Large triangle = _____ sq. cm

   Medium triangle = _____ sq. cm

   Small triangle = _____ sq. cm

   Square = _____ sq. cm

   Parallelogram = _____ sq. cm

   The area of the whole tangram puzzle is _____ square centimeters.

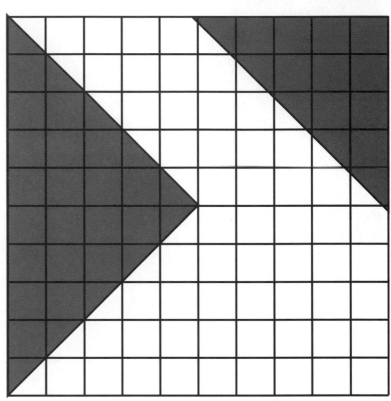

2. Use four tangram pieces to make a rectangle with an area of 50 square centimeters.

3. Use two tangram pieces to make a trapezoid with an area less than 20 square centimeters.

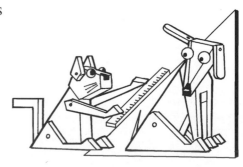

Geometry and Fractions with Tangrams
©Learning Resources, Inc.

Centimeters of Area

1. Make a pentagon with an area between 50 and 60 square centimeters. Write an addition equation to show the total amount of area.

2. Make a parallelogram with an area between 70 and 80 square centimeters. Write an addition equation to show the total amount of area.

Perimeter in Inches

Use the inch grid and an inch ruler to find the perimeter of each tangram piece to the nearest $\frac{1}{8}$ of an inch.

Place the tangram pieces as shown on the grid at the right. Determine the lengths of the tangram pieces along the outside edges of this grid.

Label the sides of each tangram piece in inches. Write an addition equation to find the perimeter of each tangram piece.

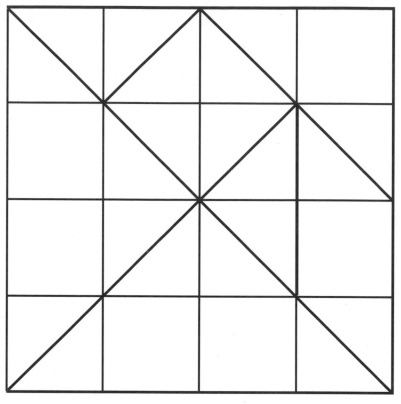

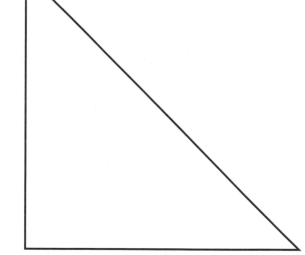

P = _____ + _____ + _____ = _____

P = _____ + _____ + _____

+_____ = _____

P = _____ + _____

+_____ = _____

P = _____ + _____ + _____ = _____

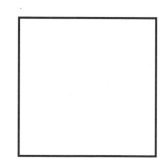

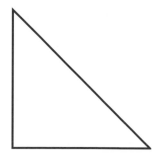

P = _____ + _____ + _____ + _____ = _____

Geometry and Fractions with Tangrams
©Learning Resources, Inc.

Perimeter in Inches

1. Use four tangram pieces to make this square. What is the perimeter? _____ inches

2. Double the lengths of each side of each tangram piece in the rectangle.

3. Use three tangram pieces to make a trapezoid with a perimeter of $10\frac{1}{8}$ inches.

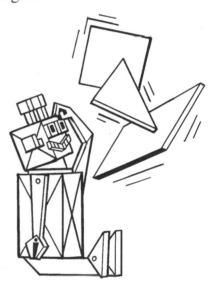

Perimeter in Centimeters

Use the centimeter grid and a centimeter ruler to find the perimeter of each tangram piece to the nearest $\frac{1}{2}$ (0.5) of a centimeter.

Place the tangram pieces as shown on the grid at the right. Determine the lengths of the tangram pieces along the outside edges of this grid.

Label the sides of each tangram piece in centimeters. Write an addition equation to find the perimeter of each tangram piece.

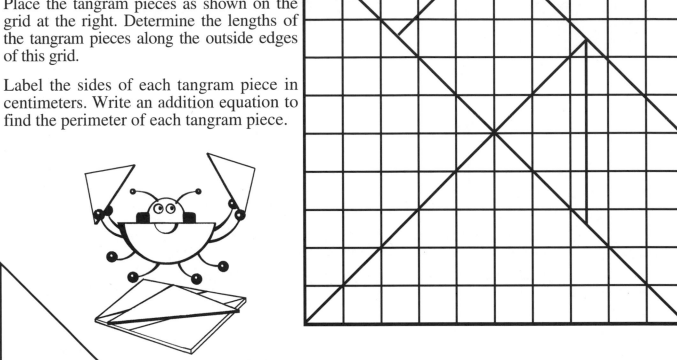

P = _____ + _____ + _____ = _____

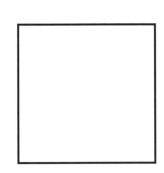

P = _____ + _____+ _____
+ _____ = _____

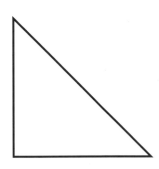

P = _____ + _____
+ _____ = _____

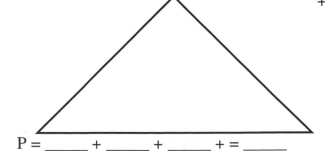

P = _____ + _____ + _____ + = _____

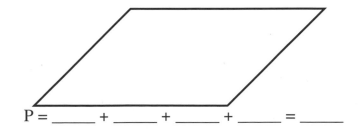

P = _____ + _____ + _____ + _____ = _____

Geometry and Fractions with Tangrams
©Learning Resources, Inc.

Perimeter in Centimeters

1. Use the medium triangle, two small triangles, a square, and a parallelogram to make a square. The perimeter is _____ centimeters.

2. Use five tangram pieces to make a hexagon with a perimeter of 27 centimeters.

3. Use four tangram pieces to make a triangle with a perimeter of 34 centimeters.

Selected Solutions

Note: At least one solution is provided for each problem or exercise; many others are possible.

Tangram Pieces and Other Shapes

Page 10: 1. triangle; 3, 3   2. triangle; 3, 3; 2   3. triangle; 3, 3; 2; 4   4. shape   5. size

Page 11: 1. square; 4, 4; equal, 2   2. parallelogram; 4, 4; 2   3. 4 sides, 4 corners   4. different size angles
5. square, parallelogram   6. 4,4   7. A, C

Page 12: 1. 90°   2. right, acute   3. no, acute, obtuse   4. 180   5. 360

Page 13: 1. square; 4, 4; the same length; congruent; right   2. parallelogram; quadrilateral; less, greater
3. triangles; 3, 3; right; acute; right triangle

Page 14:

1.

2.

Page 15:

1.   2.   3.

Page 16:

1.   2.

Page 17:

1.   2.

Page 18:

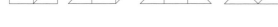

Page 19:

Geometry and Fractions with Tangrams
©Learning Resources, Inc.

Congruent and Similar Shapes

Page 22:

1.    2.   3. yes

Page 23: 1. yes   2.

Page 24: 1. similar,   2. similar,   3. yes, B; yes, A, C

Page 25: 1.   2.   3.   4. a. congruent, b. similar, c. similar

Page 26: 1.   2.   3.   4.   5. a. similar, b. congruent, c. congruent,
d. similar, e. similar, f. similar

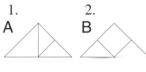

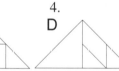

Page 27:   1.   2.   3.   4.   5. a. true, b. false, c. true, d. false,
e. false, f. false

Page 28: 1.   2. yes, answers vary   3. no

Page 29: 1.   2.   3. no   4. yes

Page 30: 1.   2. yes, 3 hexagons should be congruent
3. no
4. no

Symmetry of Tangram Shapes

Page 32:

Page 33:

Page 34:

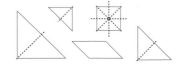

Page 35: See Teaching Notes on page 31.

Page 36: See Teaching Notes on page 31.

Area Measurement with Nonstandard Units

Page 38: 1. 2   2. 4   3. 2   4. 6   5. $\frac{1}{2}$ 6. $1\frac{1}{2}$

Page 39: 1. 2   2. 2   3. 2   4. 4   5. 5   6. 8   7. 6

Page 40: 1. $1\frac{1}{2}$ 2. 2   3. 2   4. 2   5. 3   6. 6

Page 41: Values: $500, $1000, $500, $500; $4,000   1. $1,750   2. $2,000   3.

Page 42: Values: $500, $2,000, $1,000, $1,000; $8,000   1. $4,000   2. $2,500   3.

Fractional Parts of Tangram Shapes

Page 44: 1. $\frac{1}{4}$, $\frac{1}{8}$, $\frac{1}{16}$, $\frac{1}{8}$, $\frac{1}{8}$ 2. See Teaching Notes on page 43   3. $\frac{1}{2}$ 4. $\frac{1}{4}$ 5. $\frac{1}{2}$ 6. $\frac{1}{8}$ 7. $\frac{1}{16}$ 8. $\frac{1}{8}$ 9. $\frac{3}{4}$

Page 45: 1. $\frac{1}{8} + \frac{1}{16} + \frac{1}{16} = \frac{4}{16} = \frac{1}{4}$ 2. $\frac{1}{8} + \frac{1}{16} = \frac{3}{16}$ 3. $\frac{1}{4} + \frac{1}{4} + \frac{1}{8} + \frac{1}{16} = \frac{11}{16}$ 4. $\frac{1}{8} + \frac{1}{8} + \frac{1}{16} = \frac{5}{16}$

Page 46: 1. $\frac{1}{8} + \frac{1}{8} + \frac{1}{8} + \frac{1}{16} + \frac{1}{16} = \frac{8}{16} = \frac{1}{2}$ 2. $\frac{1}{4} + \frac{1}{4} + \frac{1}{8} + \frac{1}{8} + \frac{1}{16} + \frac{1}{16} = \frac{14}{16} = \frac{7}{8}$ 3. $\frac{1}{16} + \frac{1}{16} + \frac{1}{8} + \frac{1}{8} = \frac{6}{16} = \frac{3}{8}$

Page 47: 1. $\frac{1}{4}$ 2. $\frac{3}{8}$ 3. $\frac{1}{2}$ 4. $\frac{3}{4}$

Page 48: 1. $4 \times \frac{1}{16} = \frac{4}{16} = \frac{1}{4}$ 2. $4 \times \frac{1}{16} = \frac{4}{16} = \frac{1}{4}$ 3. $3 \times \frac{1}{16} = \frac{3}{16}$ 4. $6 \times \frac{1}{16} = \frac{6}{16} = \frac{3}{8}$ 5. $2 \times \frac{1}{16} = \frac{2}{16} = \frac{1}{8}$

Page 49: 1. $\frac{1}{8} \div \frac{1}{16} = 2$ 2. $\frac{1}{4} \div \frac{1}{8} = 2$ 3. $\frac{3}{8} \div \frac{1}{16} = 6$ 4. 4   5. 2   6. 2   7. 4   8. 6   9. 6   10. $1\frac{1}{2}$

Page 50: 1. $\frac{1}{2}$, $\frac{1}{4}$, $\frac{1}{8}$, $\frac{1}{4}$, $\frac{1}{4}$ 2. $\frac{1}{3}$, $\frac{1}{6}$, $\frac{1}{12}$, $\frac{1}{6}$, $\frac{1}{6}$

Geometry and Fractions with Tangrams
©Learning Resources, Inc.

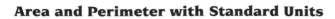

Area and Perimeter with Standard Units

Page 52: 1. 4, 2, 1, 2, 2; 16   2.   3.

Page 53: 1.   2.   3.

Page 54: 1. 25, $12\frac{1}{2}$, $6\frac{1}{4}$, $12\frac{1}{2}$, $12\frac{1}{2}$; 100.   2.   3.

Page 55: 1. $25 + 12\frac{1}{2} + 12\frac{1}{2} + 6\frac{1}{4} = 56\frac{1}{4}$ 2. $25 + 25 + 12\frac{1}{2} + 6\frac{1}{4} + 6\frac{1}{4} = 75$

Page 56: $P = 4 + 2\frac{3}{4} + 2\frac{3}{4} = 9\frac{1}{2}$ $P = 1\frac{3}{8} + 1\frac{3}{8} + 1\frac{3}{8} + 1\frac{3}{8} = 5\frac{1}{2}$ $P = 1\frac{3}{8} + 1\frac{3}{8} + 2 = 4\frac{3}{4}$

$P = 2\frac{3}{4} + 2 + 2 = 6\frac{3}{4}$ $P = 2 + 1\frac{3}{8} + 2 + 1\frac{3}{8} = 6\frac{3}{4}$

Page 57: 1. 11   2.   3.

Page 58: $P = 10 + 7 + 7 = 24$ $P = 3.5 + 3.5 + 3.5 + 3.5 = 14$ $P = 5 + 3.5 + 3.5 = 14$

$P = 7 + 5 + 5 = 17$ $P = 5 + 3.5 + 5 + 3.5 = 17$

Page 59: 1. 28   2.   3.

Terrific Tangram

Award Certificate

To: _____

For excellent work in
solving geometry and fraction
tangram activities.

Date_____   Teacher_____

Activity: _____

_____

_____

_____

_____

_____